STRENGTH IN NUMBERS: ANIMAL GROUPS

MAMMAL GROUPS

BY CLARA MACCARALD

Cover Image: Female elephants and their calves live in family groups called herds. Herd members work together to take care of their young.

Core Library
An Imprint of Abdo Publishing
abdobooks.com

abdobooks.com

Published by Abdo Publishing, a division of ABDO, PO Box 398166, Minneapolis, Minnesota 55439.

Printed in the United States of America, North Mankato, Minnesota.
052025
092025

THIS BOOK CONTAINS RECYCLED MATERIALS

Cover Photo: Henk Bogaard/Shutterstock Images
Interior Photos: Vinnie Lauritsen/Shutterstock Images, 4–5, 45; Patrick J. Endres/Corbis Documentary/Getty Images, 7; John Morrison/E+/Getty Images, 8; Jurgen & Christine Sohns/imageBROKER/Getty Images, 9; JohnnyGreig/E+/Getty Images, 12–13; Red Line Editorial, 15 (map), 25; Shutterstock Images, 15 (wildebeest), 22, 34–35, 39, 43; Philipp Rohner/500px/Getty Images, 16; KenCanning/E+/Getty Images, 19; Gerard Soury/The Image Bank/Getty Images, 20–21; Sergio Pitamitz/VWPics/Universal Images Group/Getty Images, 26; Rick & Nora Bowers/Alamy, 28–29; Photoshot/Avalon.red/Alamy, 30; Jeff Foott/Photodisc/Getty Images, 36; Serge Melesan/500px/Getty Images, 38

Editor: Laura Stickney
Series Designer: Ryan Gale

Library of Congress Control Number: 2024949008

Publisher's Cataloging-in-Publication Data

Names: MacCarald, Clara, author.
Title: Mammal groups / by Clara MacCarald
Description: Minneapolis, Minnesota: Abdo Publishing, 2026 | Series: Strength in numbers: animal groups | Includes online resources and index.
Identifiers: ISBN 9781098297282 (lib. bdg.) | ISBN 9798384919803 (ebook)
Subjects: LCSH: Mammals--Juvenile literature. | Mammal communities--Juvenile literature. | Animal colonies--Juvenile literature. | Mammals--Behavior--Juvenile literature.
Classification: DDC 599--dc23

CONTENTS

CHAPTER ONE

A WOLF HUNT

A pack of eight gray wolves trots tirelessly through dry grass. Hills and mountains rise in the distance, topped with evergreen trees. A river snakes through the landscape. The wolves live in Yellowstone National Park. This park is located in Wyoming, Montana, and Idaho. Like many wolf packs, this pack is made up of a male and female breeding pair and their offspring. The hungry wolves are on the hunt. Herds of bison wander and graze in the valley. Bison are

Most gray wolf packs include six to ten wolves, but some packs are larger. Each wolf plays a specific role in the pack.

RETURN OF THE PREDATOR

The US government created Yellowstone National Park in 1872. People wanted to protect prey animals in the area, so they poisoned and shot predators that lived in the park. In 1926, people killed the last Yellowstone wolf pack. With these predators gone, elk numbers exploded. But the huge elk population ate too many plants. Because of this, hundreds of elk starved. In 1995, people began bringing wolves back to the park. At the beginning of 2024, about 124 wolves and 10 wolf packs lived in Yellowstone National Park. Scientists have seen signs that the return of wolves may be helping to keep the elk population under control.

the largest and most deadly prey animals in Yellowstone. A bigger wolf pack might be able to take one down. But these wolves ignore the bison, looking for easier prey.

The sun is low in the sky. It lights up the undersides of clouds and casts hillsides into shadow. Finally, the dominant female wolf spies a group of elk along a wooded ridge. She starts running toward them. The rest of the pack catches her excitement and follows.

Gray wolves work together to hunt large prey such as moose. The wolves surround the animal on all sides.

A single wolf would not be able to take down an elk. But by working together, the pack can successfully hunt these large animals. The elk herd grazes peacefully. Suddenly, one elk notices the approaching predators. She runs a few steps away, hoofs stomping over the grass. But then she turns and faces the wolves again. The other elk follow her gaze. There are too many wolves for the elk to risk fighting back. The whole herd breaks into a run. Older females shield the younger elk with their bodies.

Officials at Yellowstone National Park keep track of gray wolf packs that live and hunt inside the park. One is called the Wapiti Lake Pack.

The wolf pack races behind the elk herd. Most hunts end in defeat, but the wolves are determined. They scan the elk for weaknesses. One elk is a bull. It's bigger than the others. But it's also skinny. The mating season ended a few weeks ago. During this time, bulls fight each other for female mates. The skinny bull is exhausted from fighting all fall. The wolves move in, blocking the bull off from the rest of

Members of a gray wolf pack work together to care for pups. Most pups begin going on hunts with the pack when they are six months old.

the herd. The wolves surround their prey, their tails wagging. The younger wolves stay back while the older wolves attack, slashing at the elk's legs and side. The injured elk starts to move slowly. But the wolves bite deep, finally bringing their prey down to the ground where they will feast. Thanks to living in a group, the wolves will have food for days.

MAMMALS BY THE NUMBERS

Like wolves, many mammals find strength in numbers. Mammals are vertebrates that produce milk to nurse their young. Mammals usually have fur or hair. Most give

MAMMAL ROYALTY

Naked mole rats live in eastern Africa. They live in colonies in underground burrows. These groups can include almost 300 individuals. Naked mole rat colonies have an unusual structure. The colonies have a queen, similar to honeybees. The naked mole rat queen gives birth to all the babies. The other naked mole rats are workers. They dig tunnels, gather food, and raise babies. When a naked mole rat queen dies, other females might fight each other for the chance to become queen.

birth to live young. Some mammals live alone, while others live in pairs. Many live in larger groups. Some species of mammals form groups that number in the millions.

Sometimes these animal groups are made up of unrelated individuals. Other times, families may stay together. Mothers and sisters may raise their young together. Adults of different sexes may live apart. One or a few males may live with a group of females. One animal may be the leader. Other times, individuals may be equal. Sometimes larger communities are made up of smaller family groups.

Like elk, some mammals live together because it is safer than living alone. A group has more eyes to watch out for predators. Each group member has a smaller chance of being eaten than an animal living alone. Group members can also learn from watching each other. Living together can help mammals find food. Some mammals cooperate, such as gray wolves that hunt together. Groups of predators can kill bigger prey than one animal could by itself. Living in large numbers helps many mammals survive and reproduce.

EXPLORE ONLINE

Chapter One talks about wolf packs. The website below goes into more depth on this topic. As you know, every source is different. What information does the website give about wolf packs? How is the information from the website the same as the information in Chapter One? What new information did you learn from the website?

WOLF

abdocorelibrary.com/mammal-groups

CHAPTER TWO

COMMON WILDEBEESTS

Wildebeests belong to the antelope family. They are also known as gnus. The largest wildebeests can grow up to 4.5 feet (1.4 m) tall and weigh 550 pounds (250 kg).

Wildebeests live in plains in southern and eastern Africa. They live in woodland areas in these regions, too. These mammals are herbivores. They travel throughout the year searching for water and fresh grass to eat.

The yearly wildebeest migration is often called the Great Migration. Wildebeests gather in large numbers, traveling north to the Masai Mara National Reserve.

RECOVERED HERDS

In the late 1800s, the Serengeti's wildebeests faced a huge threat. Livestock in the area started dying from a disease called rinderpest. The disease spread to wildebeests and other mammals. By the mid-1900s, wildebeest numbers in the Serengeti had dropped to about 300,000. Because so many herbivores died, more plants grew. Dry grasses and other plants fueled huge fires. The fires killed many trees. Humans created a vaccine to defend against rinderpest in the 1960s. As humans got rid of the disease in livestock, wildebeest populations grew. Fires decreased, and trees recovered.

More than a million wildebeests live in the Serengeti, a region in Tanzania that covers about 11,600 square miles (30,000 sq km). During the wet season, wildebeests live in southern plains within their range. When the dry season comes, the plains dry out. But the woodlands to the north stay wet. The wildebeests begin a long trek to the north in May or June. Other animals migrate along with the wildebeests. These migrating herds

WILDEBEEST
MIGRATION

This map of Serengeti National Park shows the routes taken by migrating wildebeest herds throughout the year. What do you notice about these routes? Think about what you've read about wildebeests. What does this map tell you about wildebeest herds? What are the herds doing at different places on the map?

Wildebeests travel in tightly packed lines and clusters during migration. At river crossings, large numbers of wildebeests leap into the water.

include other herbivores such as gazelles, antelopes, and zebras. In around November and December, the herbivores head back south. The whole journey is a huge 1,250-mile (2,000-km) loop. It is the largest migration of land mammals in the world.

HERD LIVING

Wildebeest herds change over time. When they aren't migrating, wildebeests gather in small herds. These are made up of adult females, their calves, and a bull. The breeding season starts as early as April. The male mates with females in his herd. He defends the group from other males.

During migration, wildebeests form massive herds. Long lines stretch as far as the eye can see. Over the years, these huge numbers of animals have made trails that are easy to walk on. The smell of the animals in front guides the animals in the rear. During migration, wildebeests

MIGRATION IN A MODERN WORLD

Mammal migration has become more difficult over hundreds of years. In the early 1800s, tens of millions of American bison migrated across the North American grasslands. But by the 1880s, white settlers had hunted bison almost to extinction. Herds of other mammals, such as saiga antelope in central Asia, have become much smaller too. The modern world has less room for mammals to travel. People have destroyed a lot of wildlife habitats. Some have built fences that block migrating mammals. Animals also risk getting hit by cars when they try to cross roads.

must travel over lands full of predators such as lions and hyenas. Wildebeests must also cross rivers that contain huge Nile crocodiles.

Whether staying in one place or migrating, living in large numbers protects wildebeests from predators. A large group has more animals to watch for danger. This means each individual can spend more time grazing and less time looking around. A wildebeest also has a better chance of escaping a predator if it is in a group. The predator might choose to go after a different individual. Herds can also frighten off predators. Wildebeests might stamp their feet and bunch together. Sometimes a wildebeest walks toward a predator or follows it. Female wildebeests will fight other animals to protect their calves.

A FLOOD OF CALVES

Wildebeests don't just live and migrate together. They also give birth together. Calving in the Serengeti usually happens between January and March. About 80 to

During the calving season, thousands of wildebeests are born. Wildebeest calves can stand and run shortly after being born.

90 percent of pregnant wildebeests have their calves within two or three weeks of each other. They flood the grasslands with hundreds of thousands of babies.

The huge numbers of calves help keep the young wildebeests safe. Calves are easy prey. With so many calves born at one time, predators will only be able to eat a limited number before the animals grow older and stronger. The few calves born at different times of the year struggle to survive because they are easy prey. Even baby wildebeests need large numbers to survive.

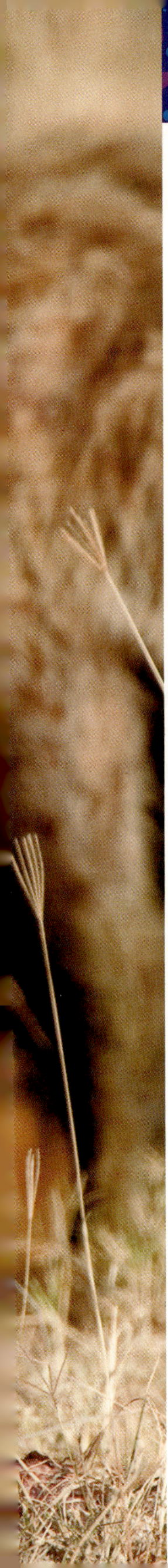

CHAPTER THREE

SPOTTED HYENAS

The spotted hyena is a social predator found throughout Africa. Many live south of the Sahara desert. These large mammals can weigh up to 176 pounds (80 kg). Although they look similar to dogs, hyenas are more closely related to cats. Hyenas are the most abundant of the large predators within their range. They hunt for food but also scavenge dead animals.

Spotted hyenas live in groups called clans. Small clans might contain only a

A group of hyenas is sometimes called a cackle. This is because hyenas make noises that sound like laughter.

Hyenas work together to protect their food and territory from animals such as lions. Lions often try to steal a hyena clan's kills or attack its cubs.

handful of individuals. Larger clans can have as many as 90 hyenas. Clans include related females and their offspring. They also include adult males who came from other clans. In general, adult female spotted hyenas are dominant over males. Hyena clans work together to protect their territory, to find food, and to protect that food from other predators.

CLAN LIVING

Hyenas need territories with a lot of prey. Some territories stay the same all year. In other places, hyenas change the shape of their home range as herbivores migrate. A clan protects its food source by keeping other hyenas out. Hyenas guard the boundaries of their land with patrols, which are often led by females. They also mark the boundaries with scents. For example, hyenas might mark a spot by leaving bodily fluids there. Other hyenas pick up these scents. If strangers invade

TO GATHER OR NOT TO GATHER

Not all hyenas live in groups. The striped hyena is a species found from northern Africa to India. A striped hyena weighs only half as much as a big spotted hyena. Striped hyenas eat meat, fruits, and insects. In some places, adult striped hyenas live alone or with young cubs. But in other places, these hyenas form groups. One female lives with a few males. The difference may be that these areas have enough food and water to support more animals. Striped hyenas benefit from staying together in these places.

PREDATOR MIGRATION

Spotted hyenas aren't the only predators who change their territories to chase migrating prey. Some gray wolves do too. In Mongolia, gray wolves prey on kulan, or Mongolian wild asses. These herbivores can migrate tens of thousands of miles every year. In 2019, scientists wanted to measure just how far the wolves and their prey traveled. They put special collars on the wolves and kulan to track their movements. The wolves not only chased after the migrating kulan but covered more ground over the year as they tried to locate prey.

the area, the hyenas attack. Hyenas make a whooping noise that calls members of the clan together to fight. If one clan seems weak, females from a neighboring clan might band together to try to take over its territory. Sometimes neighboring clans have ongoing quarrels called wars.

Female hyenas give birth alone. But when the cubs get older, the females bring their cubs together in a large den. Other hyenas in the clan live there too. In the den, the cubs start learning about their group. Within a clan,

KILLED BY HYENAS

In a 2010 study, scientists looked at two regions of Etosha National Park in Namibia. They recorded which species were killed by hyenas. Zebras and kudu are large prey animals, while springbok are smaller. Clans throughout the park numbered from 11 to 30 hyenas. These pie charts show which percentage each species made of the hyenas' total kills in an area. What do you notice about the two charts? Why do you think hyena diets in the two regions were different?

Central Etosha

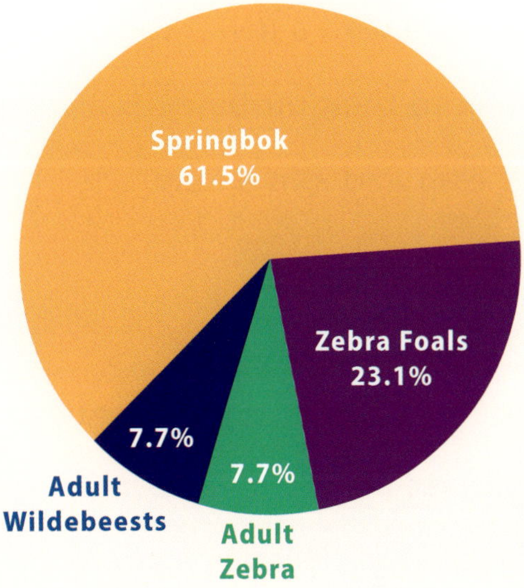

Eastern Etosha

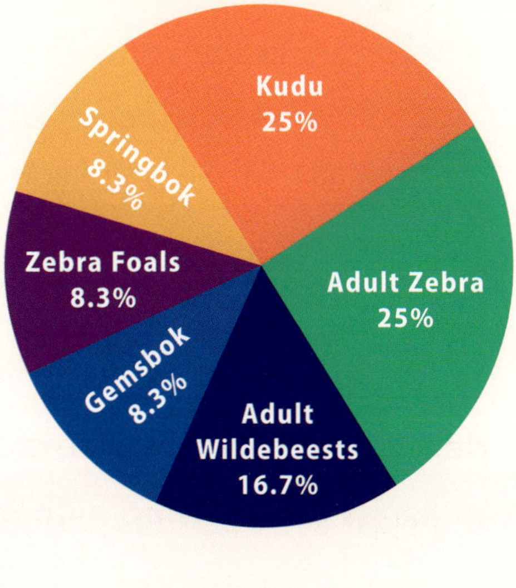

While hunting, hyenas use noises such as whoops, shrieks, or cackles. These noises may be used to warn clan members of danger or let them know a kill has happened.

individuals have different ranks. A hyena with a higher rank gets to eat more food than one with a lower rank. Hyena cubs take on a similar rank to their mothers. But males who leave their mother's clan and join a new group start at the bottom of the rankings.

ON THE HUNT

A clan doesn't always hunt together. Lone hyenas can kill and eat small prey, such as monkeys. Hyenas form larger groups to go after larger prey. A large group has an easier time driving an herbivore away from its herd.

The hunters work together to chase the prey until it is exhausted and easy to kill. Together, many hyenas can kill a big zebra or even a giraffe.

Large groups also help hyenas defend their food from other predators. Lions are large and strong. One lion can take on several hyenas. Lions can hurt or even kill the smaller animals. Hyenas use whooping calls to bring in more clan members to defend their food. A large group of hyenas can also steal kills from lions. Gathering in large numbers helps hyenas stay safe and well-fed.

FURTHER EVIDENCE

Chapter Three discusses the spotted hyena and its social life. What was one of the main points of this chapter? What evidence is included to support this point? Read the article at the website below. Does the information on the website support the main point of the chapter? Does it present new evidence?

HYENA

abdocorelibrary.com/mammal-groups

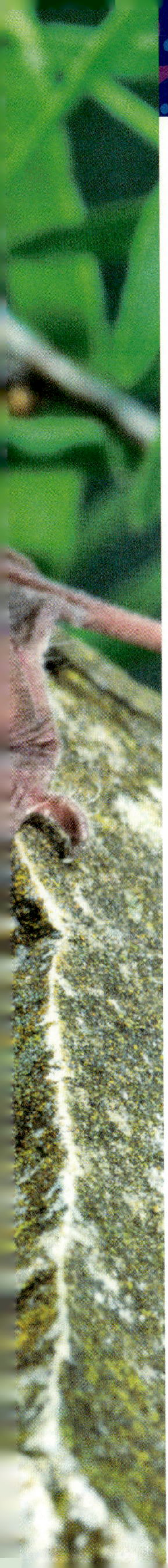

CHAPTER FOUR

MEXICAN FREE-TAILED BATS

Mexican free-tailed bats form the largest-known bat colony in the world. Bracken Cave in Texas hosts as many as 20 million bats during the summer. The species' range stretches from the southern and western United States to South America.

Mexican free-tailed bats weigh about 0.4 ounces (11 g). But their wings stretch a foot (0.3 m) or more. The bats mostly dine on moths but also eat beetles and other insects.

Mexican free-tailed bats have large ears that help them pick up many sounds. The bats use sounds to communicate too. These may be used to identify other bats or to locate pups.

When Mexican free-tailed bats fly out of Bracken Cave, they form a large, swirling spiral. Many people call this formation a "batnado."

Some populations of Mexican free-tailed bats migrate. They head south in the fall to spend the winter in warmer locations. In the spring, they return north to have their babies. Every year, many bats return to the same locations. Over time, guano builds up underneath the colony. Huge piles of rotting guano make the air toxic inside the caves that house large bat colonies. The bats can handle the chemicals. But their predators cannot, so they do not invade the cave. This is one way that Mexican free-tailed bats benefit from living in large numbers.

IN THE COLONY

Bats roost together during the day and at night. In cold weather, roosting together helps bats stay warm. Otherwise, they could get too cold and die. Many bat colonies roost in caves. Others form under bridges or in buildings. At night, bats emerge to find food. With so many hungry mouths, the bats can't eat too close to their colony. Instead, they travel over a wide area. They stick together in small groups while feeding.

In summer, Mexican free-tailed bats separate by sex. Males form their own colonies. Pregnant females gather in huge groups. The mothers give birth to one to three pups. When the mothers go out to feed, they

BAT RADAR

Weather radar works by detecting objects in the air. It sends out signals and listens for echoes, which indicate that the signal hit something. For example, large areas of rain in the air show up on a weather map. But weather radar can also detect flying animals. When millions of bats emerge from a colony such as the one at Bracken Cave, the animals can appear on weather maps.

BLOOD FRIENDS

Female vampire bats feed pups. But they also help other adults. A bat friend can save a life. There are three species of vampire bats, all of which live on a diet of blood. Small groups of female vampire bats roost together for many years in large colonies. Vampire bats must feed every few nights or they will starve to death. That's where a bat's friends come in. A vampire bat that got a blood meal can share some of the blood with a friend who wasn't so lucky. It does this by throwing up some of the blood it ate.

leave their pups in large groups.

Bat pups can't stay warm by themselves. But groups of pups can produce heat by huddling together. This mass of babies makes it hard for a mother bat to find her baby. Mothers and babies use sound and smell to recognize each other. About eight out of ten times, a mother nurses her own pup. But the rest of the time, mothers nurse unrelated pups. Pups also form bonds. A female pup will likely grow up to give birth in the same cave she was born in. Older females teach her where to roost and feed.

STRAIGHT TO THE SOURCE

Fran Hutchins works for Bat Conservation International. He spoke about the feeding patterns of Mexican free-tailed bats roosting in Bracken Cave:

> *These bats [eat] their body weight in bugs, which are primarily agricultural pests, . . . traveling about 60 miles [97 km] one way to forage over where we have our corn and cotton and sorghum crops. They're going to fly 60 miles [97 km] from here, and up to 10,000 feet [3,048 m] in altitude, depending on where those large insect swarms are. They're going to eat their body weight in bugs, which is over 150 tons [136 metric tons] of bugs tonight. And they're going to come back, nurse their young, and do this all over again tomorrow.*

Source: Jack Morgan. "Batnado at Bracken Cave Is an Incredible Natural Phenomenon with 20 Million Players." *Texas Public Radio*, 16 July 2024, tpr.org. Accessed 28 Aug. 2024.

CONSIDER YOUR AUDIENCE

Adapt this passage for a different audience, such as your principal or friends. Write a blog post conveying this same information for the new audience. How does your post differ from the original text and why?

CHAPTER FIVE

ORCAS

The orca, also known as the killer whale, is the largest member of the dolphin family. Males can measure up to 30 feet (9 m) long. Orcas are predators that hunt fish and other mammals, including whales.

Orcas are a wide-ranging species that live in every ocean on Earth. In some populations, orcas live in small territories. In others, orcas travel widely across large territories.

Orca pods often include multiple generations and are made up of several family units. The pods are matriarchal, which means females are the leaders.

Female orcas give birth to one calf at a time. The calf usually stays with its mother for its whole life, becoming part of the mother's pod.

Many orcas live in groups called pods. Orca pods can contain up to 50 individuals, although many have 20 or fewer. Pod members are often related to each other on the female side. Calves stay with their mothers to nurse for at least a year. The mothers continue to help their calves for at least a few years after that. An orca may still get support from its mother 30 years later.

Orcas are very intelligent and social. Groups eat, travel, and rest together. Pod members communicate with each other. Orcas learn from each other too. Different groups of orcas have their own cultures.

A pod's culture affects how an orca communicates. It also affects what an orca eats and how it hunts.

MAMMAL CULTURE

Many other mammals have cultures. For Japanese macaques, or snow monkeys, group culture can determine whether the monkeys soak in hot springs or wash their food. Chimpanzee culture can include cracking nuts with stones. Some chimpanzees use sticks as tools to gather and eat termites. Different groups make tools in different ways. Bottlenose dolphin culture includes tool use as well. A group of dolphins in Australia uses large shells to catch fish.

WHALE TALK

Orcas have a lot of things to say to each other. Sometimes they use their bodies to say things. When an orca is angry, it might bang its head into another whale. Sound is important too. Orcas make many different whistles, clicks, and other noises. Some calls help pods stay together. Other sounds help the animals cooperate.

Communication methods are part of an orca pod's culture. Each pod has a specific set of sounds that

Orcas use echolocation to navigate. This involves making sounds and listening to their echoes, which helps an orca determine how big an object is or how far away it is.

its members learn to use. This is known as a dialect. Dialects help group members bond with each other. Several pods can share a female ancestor. These pods will also share calls. Together, the pods create larger groups called clans. Shared calls within a clan may be used to help orcas locate each other.

Some orcas have been seen working together to catch seals by beaching themselves on shorelines. This scares the seals into the water, where other orcas can catch them.

WOLVES OF THE SEA

Living together helps orcas find food. Group members teach orca calves what animals to eat. What to eat and how to hunt are part of their culture. Orcas are large and strong. A single orca can kill a great white shark. But orcas often hunt in a group. They use different hunting strategies. Group members can keep a shark from escaping while another orca attacks. One orca can make a wave that sweeps a seal into the water, right into the mouth of another waiting orca. Groups of orcas

ORCA MENOPAUSE

Menopause is a time in a female animal's life when it stops being able to have offspring. Humans go through menopause too. However, most other mammals do not. Orcas are one of the few that do. Having calves takes a lot of energy. By going through menopause, female orcas can live decades longer than they would otherwise. Female orcas can outlive males by 20 years or more. One reason scientists think orcas go through menopause is because older females help their grandchildren and relatives survive. They couldn't do that as easily if they had calves of their own.

also herd fish into tight balls that are easier to eat. The orcas blow bubbles and slap the water to control the fish.

When food is scarce, the pod turns to its oldest female for help. Female orcas often live 60 years or longer. The oldest female will lead the pod using the knowledge gained over her long life. Group members share food and hunt together. Working together to eat and survive is just one of the many ways mammals find strength in numbers.

STRAIGHT TO THE SOURCE

Some orcas have damaged and sunk boats. Researchers are unsure why orcas behave in this way. Monika Wieland Shields works for the Orca Behavior Institute. She said:

> *I've spent more than 20 years observing wild killer whales and I've seen what they can do to their prey. . . . But when I watch the videos of these so-called attacks, I see something completely different. There's no ramming, no tearing the vessel apart. Instead it looks like another behavior we often see killer whales engage in: play. Orcas are cultural creatures. . . . Their dietary preferences, vocalizations, and travel patterns are learned from family members and make them so unique from one another. . . . And along with this societal complexity comes a unique capacity for amusement and playfulness.*

Source: Monika Wieland Shields. "I Study Killer Whales. The So-Called 'Orca Uprising' Is Just a Misunderstanding." *MSNBC.com*, 17 May 2024. Accessed 23 Sept. 2024.

BACK IT UP

The author of this passage is using evidence to support a point. Write a paragraph describing the point the author is making. Then write down two or three pieces of evidence the author uses to make the point.

FAST FACTS

- Mammals are vertebrates that produce milk to nurse their young. Many mammals benefit from living in groups.
- Gray wolves hunt in groups. Living in large numbers helps wolves tackle larger and stronger prey.
- Every year, more than a million common wildebeests migrate through the Serengeti in massive herds. The wildebeests are part of the largest migration of land mammals in the world.
- Living in herds helps wildebeests watch out for predators. It's also easier to escape predators when traveling in a group.
- Spotted hyenas live in clans. Clan members defend their territories and hunt together.
- By working together, hyenas can protect their kills from lions. Groups of hyenas can also steal food from other predators.
- Mexican free-tailed bats form the largest known bat colony in the world.
- Roosting in large numbers keeps Mexican free-tailed bats warm. Bats that get too cold can die.

- Orcas live in groups called pods. Pods have unique cultures and dialects.
- Orcas learn different hunting methods from their pods. Orcas also work together to catch food.

STOP AND THINK

Tell the Tale

Chapter One of this book discusses a wolf hunt in Yellowstone National Park. Imagine you are seeing a wolf hunt in action. Write 200 words about what you see. How do the wolves work together to catch their prey?

Surprise Me

Chapter Five discusses orca pods. After reading this book, what two or three facts about orca pods did you find most surprising? Write a few sentences about each fact. Why did you find each fact surprising?

Take a Stand

Large groups of mammals need large spaces to live in. But many mammals are losing their habitats around the world. Some people are trying to protect natural areas and the animals who live in them. Do you think people are doing enough to help mammal groups? If not, what more do you think people should do to help? Why?

Why Do I Care?

You might not live near wolves or wildebeests. But that doesn't mean mammals don't form groups near you. Look around. What kind of mammals do you see? Do any of them spend time with other members of their species? How do you think those animals benefit from living in a group?

GLOSSARY

cooperate
to work together

culture
a set of social practices or behaviors that are shared by a group

extinction
the process of a species disappearing from Earth

guano
bird or bat droppings

habitat
a place where an animal normally lives

herbivore
an animal that eats plants

migrate
to move from one place or region to another

range
the geographic area in which an animal lives

roost
to gather together to rest or sleep

scavenge
to eat animals that are already dead

species
a group of animals that can mate and have offspring with each other

territory
an area of land that animals live in and defend from others

vertebrate
an animal with a backbone or spine

ONLINE RESOURCES

To learn more about mammal groups, visit our free resource websites below.

Core Library CONNECTION
FREE! COMMON CORE MULTIMEDIA RESOURCES

Visit **abdocorelibrary.com** or scan this QR code for free Common Core resources for teachers and students, including vetted activities, multimedia, and booklinks, for deeper subject comprehension.

Booklinks NONFICTION NETWORK
FREE! ONLINE NONFICTION RESOURCES

Visit **abdobooklinks.com** or scan this QR code for free additional online weblinks for further learning. These links are routinely monitored and updated to provide the most current information available.

LEARN MORE

Deresti Betik, Lisa. *Think Like a Goat.* Kids Can, 2023.

Hudak, Heather C. *Wolf, Coyote, and Other Packs*. Crabtree, 2023.

INDEX

About the Author

Clara MacCarald is a freelance writer with a master's degree in ecology and natural resources. She lives with her family in an off-grid house in the forests of central New York and writes nonfiction books for children.